my guide to the planets

Neptune

CHERRY LAKE PRESS

Published in the United States of America by Cherry Lake Publishing
Ann Arbor, Michigan
www.cherrylakepublishing.com

Reading Adviser: Marla Conn, MS, Ed, Literacy specialist, Read-Ability, Inc.
Book Designer: Jennifer Wahl
Illustrator: Jeff Bane

Photo Credits: ©NASA images/Shutterstock, 5, 23; ©Vadim Sadovski/Shutterstock, 7, 11, 17, 19; ©Zakharchuk/Shutterstock, 9; ©iofoto/Shutterstock, 13; ©photosoft/Shutterstock, 15; ©Fasttailwind/Shutterstock, 21; Cover, 2-3, 10, 12, 22, Jeff Bane; Various vector images throughout courtesy of Shutterstock.com/

Library of Congress Cataloging-in-Publication Data

Names: Devera, Czeena, author. | Bane, Jeff, 1957- illustrator. | Devera, Czeena. My guide to the planets.
Title: Neptune / by Czeena Devera ; illustrated by Jeff Bane.
Description: Ann Arbor, Michigan : Cherry Lake Publishing, [2020] | Series: My guide to the planets | Includes index. | Audience: K-1.
Identifiers: LCCN 2019032864 (print) | LCCN 2019032865 (ebook) | ISBN 9781534158832 (hardcover) | ISBN 9781534161139 (paperback) | ISBN 9781534159983 (adobe pdf) | ISBN 9781534162280 (ebook)
Subjects: LCSH: Neptune (Planet)--Juvenile literature.
Classification: LCC QB691 .D48 2020 (print) | LCC QB691 (ebook) | DDC 523.48--dc23
LC record available at https://lccn.loc.gov/2019032864
LC ebook record available at https://lccn.loc.gov/2019032865

Printed in the United States of America
Corporate Graphics

table of contents

About the author: Czeena Devera grew up in the red-hot heat of Arizona surrounded by books. Her childhood bedroom had built-in bookshelves that were always full. She now lives in Michigan with an even bigger library of books.

About the illustrator: Jeff Bane and his two business partners own a studio along the American River in Folsom, California, home of the 1849 Gold Rush. When Jeff's not sketching or illustrating for clients, he's either swimming or kayaking in the river to relax.

About Neptune

I'm Neptune. I'm called the blue planet.

I am a large planet. I am the fourth-largest planet.

I am the farthest planet from the Sun. It's pretty cold here.

I **orbit** around the Sun. It takes me almost 165 years to complete 1 orbit!

There are four **outer planets** in the **solar system**. I am one of them.

I am an **ice giant**. This means I'm made mostly of ice.

I have 14 **confirmed** moons.

I have six rings that surround me. These rings are made up of special rocks and dust.

I am the windiest planet in the solar system. My winds are almost as fast as a jet!

I am a **unique** planet. I'm far away. Not much is known about me. There is much more to discover!

glossary

confirmed (kuhn-FIRMD) found to be true

ice giant (ISE JYE-uhnt) a giant planet composed mainly of ice

orbit (OR-bit) to travel in a curved path around something

outer planets (OU-tur PLAN-its) the four planets that are farthest from the sun

solar system (SOH-lur SIS-tuhm) the Sun and all the things that orbit around it, like planets

unique (yoo-NEEK) the only one of its kind

index